Wooc … les:
their r … dlife

M.S. Warren and R.J. Fuller

Second edition

1993

Dr Martin Warren is now a conservation consultant but previously did in-house contract work on insects for the Nature Conservancy Council.

Dr Rob Fuller is Director of Habitats Research at the British Trust for Ornithology, The National Centre for Ornithology, The Nunnery, Thetford, Norfolk IP24 2PU, which is partly funded by the JNCC.

British Trust for Ornithology

Contents

Preface

Much of the research commissioned by the Joint Nature Conservation Committee is directed at the conservation requirements of wildlife or, if primarily designed for other purposes, attempts to make results available to provide guidelines for management purposes also. Several long-term monitoring studies, particularly on nature reserves where experimental manipulations are possible, have been designed to produce information over the time-scale necessary to examine the natural processes relevant to habitat management. This information can then be applied over much wider areas.

There is clearly a great demand for such information. In the case of woodlands, this is demonstrated by the enquiries, received by both the authors of this booklet and by the staff of JNCC and the country nature conservation agencies, from a range of people responsible for particular woodlands. This great demand was emphasised even more by the response to an advance presentation of these booklets to the British Ecological Society's 1989 Symposium on "The scientific management of temperate communities for conservation".

All these sources of enquiry make clear a need not just for a prescriptive guide or recipe book for management work, but for an explanation of studies on which guidance is based, so that reasons for the management options can be assessed. Although such studies may have been published previously in scientific journals, they may not be easily accessible in a form convenient for conservation managers. Accordingly, this booklet is based upon long-term studies commissioned by JNCC from the British Trust for Ornithology on birds, and from the Institute of Terrestrial Ecology and various ecological consultants for invertebrates, as well as in-house staff. To these has been added the expertise of colleagues in other subject areas, particularly of course woodland ecology.

This and its companion volume, *Coppiced woodlands: their management for wildlife,* are the first two of a series of booklets on aspects of management of various habitats for wildlife. I would welcome any comments or suggestions as to features which would further enhance their value to practitioners. I would also like to thank Rob Fuller and Martin Warren for the extra efforts necessary from authors when developing a new series and their aid in producing the second editions.

Dr M.W. Pienkowski
Director Life Sciences
JNCC

Introduction

The survival of many plants and animals within British woodlands depends on the existence of open areas with few trees. Since the early Middle Ages, such habitats were maintained in many lowland woods by coppicing, a process whereby portions of the wood were cut on a rotation, usually of 5 – 20 years, to provide crops of poles rather than large timber. In the nineteenth century coppicing started to give way to plantation forestry which, in most woods, led to a reduction in the area of open habitat. Nowadays, in many woods, open areas are confined to rides and glades. These features make up a relatively small part of most woods, but their value for woodland wildlife can be out of all proportion to their size, provided that they are suitably managed. This booklet explains why rides and glades are so important and gives guidelines on how best to manage them for nature conservation. The management of the surrounding trees, which is equally important for woodland conservation, is dealt with elsewhere (see Further reading).

A ride is regarded as any linear opening or track within the wood and is taken to include all the area between the trees on either side. Hence this includes the grassy surface of the ride itself, together with any ride-side ditches, and the scrub or coppice regrowth which often develops between the ditch and the trees. All these components of the ride are vitally important to wildlife in woods and taken together can support a very wide range of plants and animals. Woodland glades provide similar conditions for wildlife and are defined as all permanent non-linear areas within woods which have no or few trees. Glades are permanent features, so they do not include young plantations or clearings containing developing woodland. Our recommendations on ride and glade management apply equally to deciduous and conifer plantations and to semi-natural woods throughout Britain, unless stated otherwise.

The origin and history of woodland rides and glades

There have been open areas in woodland ever since it became the dominant vegetation cover of Britain after the last ice age. In the primeval forests, openings would have been created by trees falling down as a result of disease, decay or storms. Most of these treefall gaps would have been recolonised by trees and shrubs, but some may have been maintained as glades by wild grazing animals such as deer and the now extinct aurochs. Since Mesolithic times many clearings have been created by man. Some were cultivated and quickly abandoned but in the Neolithic and Bronze Ages a vast area of forest was cleared permanently for agriculture.

During the Middle Ages, and probably earlier, glades within woodland were kept clear either by grazing domestic stock or by cutting for hay. Furthermore, many medieval Forests contained well-defined areas of deer pasture, known as plains or lawns, in amongst tracts of woodland (Figure 1a). In private deer parks the open compartments were called 'launds'. On acid soils, some glades would have been dominated by bracken just as they are today. Such bracken glades can be more or less permanent, especially where browsing by deer is heavy, because bracken can prevent the establishment of most tree seedlings. Bracken was regularly harvested from such glades in medieval times and used for animal bedding. Most of these ancient glades have now been either replanted, usually with conifers, or converted to farmland. Others have closed over through natural invasion by scrub.

Medieval records tell us little about rides, and it seems probable that the construction of networks of permanent rides in woods began only in the 16th century. Presumably before this time woods contained just temporary tracks. Most early rides were straight, which has led to the suggestion that they were established to give a long field of fire for sporting guns, which were introduced at that time (Figures 1a and 1b). The straightness of rides helps to differentiate them from old trackways or lanes in woods, which tend to follow a more sinuous course. Another major use for rides was to provide better access within the wood, primarily for timber extraction. To provide a firm surface throughout the year, rides were often drained by ditches on either side. Rides were maintained either by grazing or by cutting for hay but few examples of these practices survive today. The majority of rides are now maintained by regular cutting, usually with a tractor-drawn swipe or flail.

Today, most woods have a network of rides which are maintained primarily to aid timber extraction. In conifer plantations, rides can also act as fire-breaks and give access for fire-fighting. In commercial woodlands, the rides are usually classified either as main rides (typically 6 – 9 m wide) or secondary rides (typically 3 – 5 m wide). With the development of large

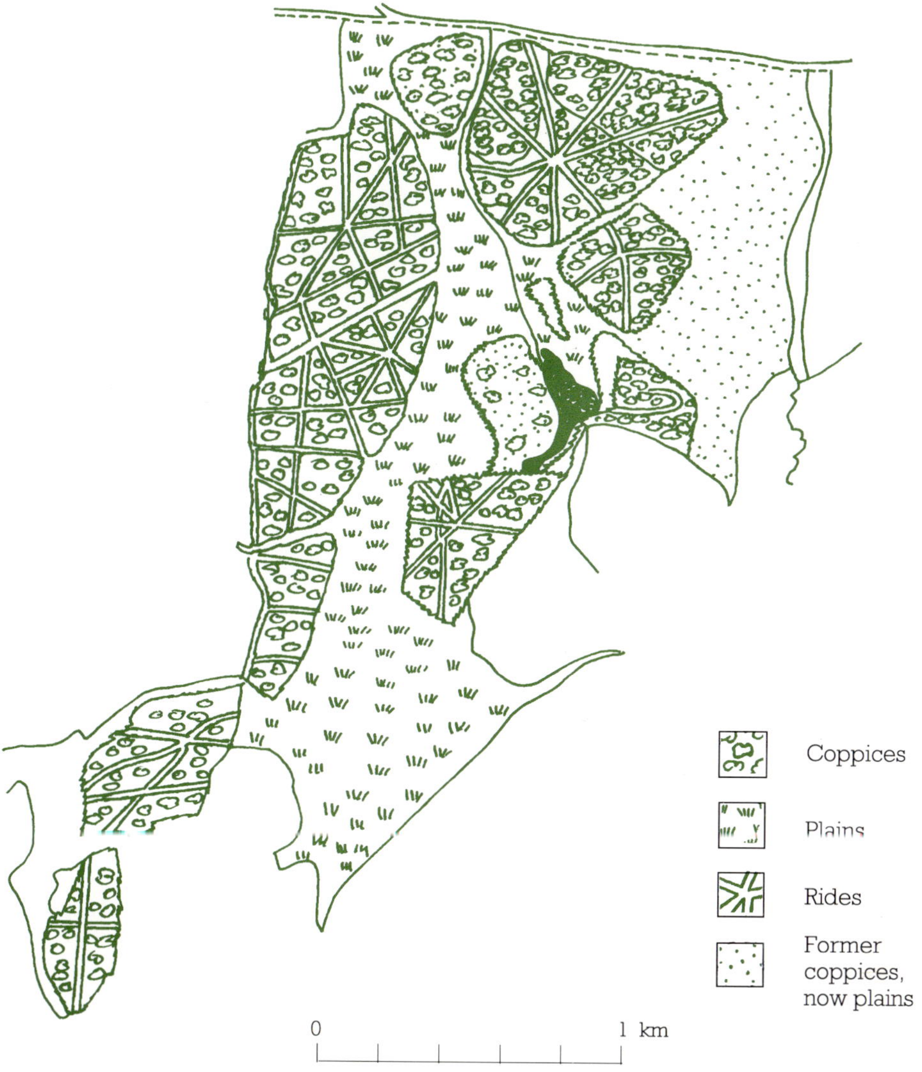

Figure 1a Example of layout of rides in an ancient wood: Hatfield Forest, Essex (after Rackham 1980)

machinery for timber extraction, selected rides have often been upgraded to form a network of woodland roads, sometimes with metalled surfaces, bordered by drainage ditches. In many cases, rides are still important for game-shooting, which can provide a valuable additional source of revenue. With the increasing demand for countryside recreation, rides have become important for leisure amenities, providing easy routes for pedestrians and horses.

Figure 1b Example of layout of rides in an ancient wood: Monks Wood National Nature Reserve, Cambridgeshire

The value of woodland rides and glades

Rides and glades tend to develop a completely different flora and fauna from the rest of the wood. Large numbers of light-demanding plants and animals thrive within rides and glades solely because they provide open, sunny conditions; such species also occur in suitable habitats outside woods. Others rely on the interface between grassland and woodland habitats, and are more or less confined to woods or wood edges. The margins of rides and glades are often particularly important in coniferous plantations, as they sometimes contain the only deciduous trees and shrubs, with their associated rich wildlife, in the whole wood. They have also become increasingly important as refuges for sun-loving plants and animals since the decline of coppicing, which was the dominant form of woodland management in lowland Britain until the 19th century. Today, only 2% of British woodland is managed as coppice, and many woods have become dark and shady. As a result, numerous species have disappeared from large parts of the country, and many more have become totally reliant on the few open rides and glades that remain within some woods. Thus, rides and glades contain a range of habitats, some of which are virtually unique, providing opportunities for a wide variety of wildlife.

Plants

The plant communities contained by rides and glades can be important relics of unimproved grassland, over 95% of which has been lost from the surrounding countryside during the last 50 years. In intensively farmed areas like East Anglia, habitat loss has largely restricted many grassland, and some wetland, plants to woodland rides and glades. Within woodland, many plants will not tolerate dense shade, but prefer more open areas, either beneath a sparse canopy or in rides and glades. Many of these plants are typical of unimproved grassland, for example bird's-foot-trefoil, yellow-rattle and many grasses. In areas of intensive agriculture, plants like saw-wort and devil's-bit scabious may survive only in the grassland still remaining in woods. Few plants are restricted to woodland rides and glades, but many are commoner there than in almost any other habitat. Examples are hedge woundwort, creeping-jenny, bitter-vetch, common figwort, wood speedwell and wood small-reed. The few specialised plants that are largely confined to rides and glades or to wood edges include wood vetch, narrow-leaved everlasting-pea, greater burnet-saxifrage and crested cow-wheat.

Those rides which are wide and sunny tend to have the greatest variety of plant species. Open grassland species, which are rare or absent in shady rides, flower more profusely in sunny conditions and cease flowering when shade becomes too dense. High numbers of plant species can be encouraged by periodic cutting, partly because this controls the more competitive species which might otherwise dominate the vegetation and partly because it provides some ground disturbance which encourages the germination of seeds. Different species will be favoured by different cutting regimes, but periodic cutting (or grazing) is usually needed to maintain the open conditions that give rides and glades their characteristic flora. Open, sunny rides which are rich in flowers also offer the greatest opportunities for insects.

Invertebrates – general

A huge number of invertebrates occur in woodland rides and glades, but, apart from butterflies and moths, comparatively little is known about their precise requirements. Most invertebrates have annual life-cycles and need suitable breeding conditions every year if populations are to survive, unlike most plants which can survive periods of adversity either as buried seed or as rootstocks. Furthermore, invertebrates frequently have completely different requirements at different stages of their life-cycles. Most plant-feeding insects are host-specific, so the greatest numbers of species tend to be found in those places with the most varied vegetation. Practically every species of grass, sedge, herb, shrub or tree has its own invertebrate fauna which feeds on it more or less exclusively. For example, the longhorn beetle *Saperda populnea* forms galls on the leaves of young aspen growing in rides, and the leafhoppers *Dicranocephalus medius* and *Oncochila simplex* both breed on wood spurge in open, sunny rides and glades. Invertebrates often show further specialisation, such as feeding only on the flower-heads of one or two plant species or only in the stems or roots of plants. Grass and sedge tussocks are important to invertebrates such as leafhoppers, beetles and bugs, not only as a source of food but also for shelter during the winter.

Bees and wasps often need patches of bare ground, partly because these provide nesting sites and partly because they provide a warm microclimate. Crumbling, south-facing banks or even rides with rutted or disturbed ground are particularly valuable. Short, sparse vegetation is important for scavenging ground beetles and certain spiders, while taller vegetation or scrub is used by other spiders and many plant-feeding invertebrates. These in turn support a range of other invertebrate predators and parasites, as well as providing food for larger animals such as birds.

The diversity of many invertebrate groups is strongly influenced by the level of shade in woodland rides, but comprehensive data are unavailable. A recent study found that open, sunny rides contained many more heteropteran bugs than shady rides, in terms both of species numbers and of abundance. Sunny rides also contained more chrysomelid beetles and seemed to be richer in weevils and certain gall flies. In contrast, leaf-mining moths feeding on hazel at ride edges showed little overall preference for the level of shade.

Flower-rich rides and glades provide vital sources of nectar and pollen to a great variety of insects, notably hoverflies and bees. These attract predators such as dragonflies and damselflies, which often hawk along rides in search of insect prey. Many invertebrates are virtually confined to the more open areas within woods, but others fly there merely to visit flowers and breed elsewhere. These include numerous insects breeding in mature woodland habitats, such as certain hoverflies and beetles whose larvae feed on dead or decaying timber.

Butterflies and moths

Woodland is a particularly important habitat for butterflies, most of which rely solely on open areas such as rides and glades. The three major features that determine the suitability of a ride for butterflies are the amount of shade, the species composition of the vegetation and the structure of the vegetation. Butterflies are sun-loving insects and the majority breed only in very open rides and glades which provide particularly warm and sunny microclimates (Table 1). Two species, the wood white and the ringlet, prefer lightly shaded rides, and only two can tolerate dense shade – the green-veined white and the speckled wood. The relative abundance of different butterfly species will change as shade levels increase or decrease, for example when surrounding trees grow or when rides are widened (Figure 2).

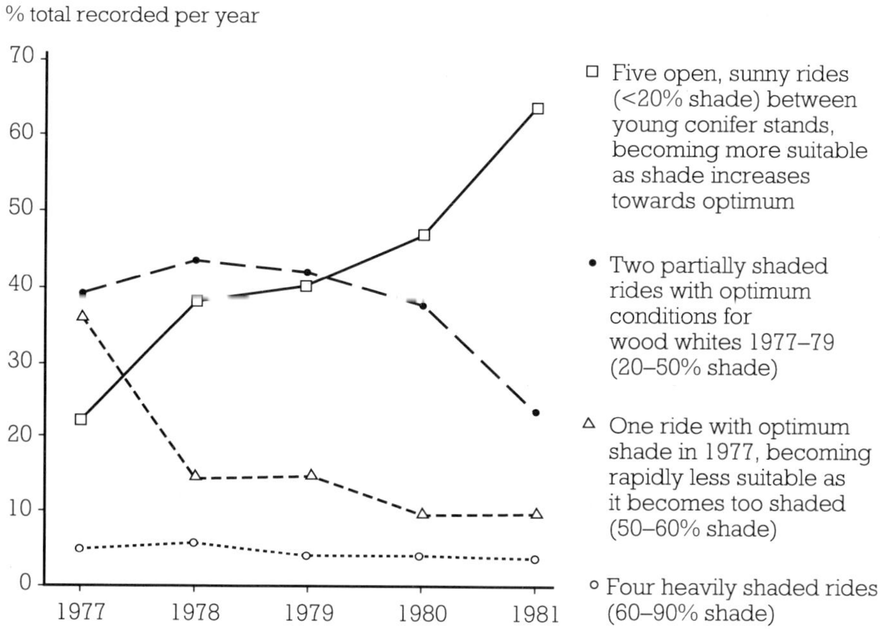

Figure 2 Yearly changes between 1977 and 1981 in the proportions of wood white butterflies recorded in woodland rides running between conifer stands of different ages (after Warren 1985)

Very open, sunny rides or glades (e.g. less than 20% shade in Figure 3)	Lightly shaded rides or glades (e.g. 10-40% shade in Figure 3)
Chequered skipper Small skipper Essex skipper Large skipper Dingy skipper Grizzled skipper * Brimstone + Orange-tip Green hairstreak + Brown hairstreak + Small copper Brown argus Common blue Holly blue + Duke of Burgundy * Painted lady Red admiral Small tortoiseshell Peacock Comma Marsh fritillary Small pearl-bordered fritillary * Dark green fritillary * Wall Scotch argus Marbled white Hedge brown Meadow brown Small heath	Ringlet Wood white **Fairly heavily shaded rides or glades (e.g. 40-90% shade in Figure 3)** Speckled wood Green-veined white **Newly cut woodland (including ride margins and coppice)** Pearl-bordered fritillary High brown fritillary Heath fritillary **Dappled shade within woodland or along edges of wood or rides** White admiral Silver-washed fritillary **Tree- or shrub-feeders, mostly confined to the canopy** Purple hairstreak White-letter hairstreak Black hairstreak + Purple emperor Large tortoiseshell

* These species also breed in newly cut woodland.
+ These species require particular trees/shrubs.

Table 1 Main habitat preferences of butterflies breeding in British woods

The species composition of the vegetation is important because the larvae (caterpillars) of each butterfly species feed on only one or two types of plant and most adults require an abundant supply of nectar. The structure of the vegetation is important because, when laying eggs, butterflies are highly selective not only about the precise type of food-plant for their larvae but also about the exact position in which it is growing. Some butterflies, such as the grizzled skipper and small heath, select short vegetation, while others, such as the wood white, marbled white and ringlet, prefer tall, grassy areas. A few butterflies prefer newly cut woodland, where larval food-plants such as violets grow in sparse vegetation. These include the pearl-bordered, high brown and heath fritillaries, which have all declined drastically in England since the widespread cessation of coppicing. By cutting ride margins on a rotation, it is possible to produce conditions similar to those created by coppicing.

The few butterflies that breed on tree and shrub species are rarely seen in rides or glades. One of these, the purple emperor, often breeds on sallows growing in sheltered ride margins. Two species, the silver-washed fritillary and white admiral, tend to breed in dappled shade beneath the woodland canopy. However, the former prefers to breed around the shady edges of glades, and the latter often lays its eggs on honeysuckle growing in scrubby ride margins. Both these and many other butterfly species rely heavily on sunny rides to obtain nectar from flowers. Woodland glades are similar to rides in the conditions they offer for butterflies, although sometimes they can provide habitats for species like the marsh fritillary, which needs reasonably large, open areas of damp grassland.

In contrast to butterflies, a much larger proportion of moths breed on tree and shrub species. The shrubby margins of rides are, therefore, vital to moths, especially in conifer plantations where broadleaved trees are otherwise scarce. Even in deciduous woods, several tree and shrub species tend to be more abundant in the ride margins or at the edges of woods. Many moths are fairly specific with regard to their food-plant, so the more varied the shrub margin, the greater the number of moth species. Sallows and aspen are particularly important ride-side trees for moths, including the poplar hawk, puss moth, sallow and poplar kittens, common wave, chocolate tip, swallow and pebble prominents and bordered beauty. They are also the sole food-plant of rarer species like the light orange underwing, which breeds on aspen. Other trees and shrubs including blackthorn, wild roses, wayfaring-tree, guelder-rose, buckthorn, wild privet, spindle and field maple have associated moths which breed exclusively on them.

The importance of the age or height of trees and shrubs to moths is poorly known, largely because of the difficulties of sampling the crowns of mature trees. However, ride-side regrowth of birch, hawthorn and other woody plants supports large numbers of moth larvae, even when only one or two metres tall. This suggests that the combination of sun and shelter from wind in rides provides highly favourable breeding conditions for many species. For example, the caterpillars of the common white wave occur at high density on low birch and the brindled white-spot can be plentiful on hazel regrowth only two or three years old. Other moths, such as many leaf-mining species, seem to prefer taller shrubs or mature trees. Consequently, the greatest number of moth species tends to occur where rides have shrub margins composed of varied ages and species of shrubs. Like the butterflies, those moths that do use low-growing plants in woodlands are fairly host-specific, and these include several rare species like the drab looper (on wood spurge) and the mere wainscot (on wood and purple small-reed). Most of these species can breed very successfully in rides and glades, provided that their larval food-plants are abundant. So, although the moths themselves may not depend as much as butterflies on the warm microclimate in rides and glades, they often rely on them because they contain their larval food-plants.

Reptiles and amphibians

Several reptile species can use open areas in woods, particularly on acidic, sandy soils. In Dorset, a detailed study in a dry conifer plantation on former heathland has shown that wide rides can provide habitats for lizards and probably snakes. The most suitable habitat for the rare sand lizard was on the banks of sunny rides where the vegetation was dominated by heather (ling), bell heather, bilberry and bracken. Banks with a southerly or easterly aspect and with bare ground were preferred, as these allowed the animals, especially young ones, to take maximum advantage of the warmth of the early morning sun. The common lizard showed a slightly weaker preference for the same aspects, but mainly occurred where the vegetation was dominated by purple moor-grass, cross-leaved heath, gorse and various grasses. For both species, high densities were associated with broad, sunny rides with wide, uncut verges. In damp broadleaved woods, rides can be valuable refuges for grass snakes and, where the ditches are permanently wet, for frogs.

Birds

The value of rides as habitats for nesting birds is critically dependent on their width and on the nature of the vegetation at the ride margin. In general, narrow rides and those with no marginal belt of shrubby vegetation are the least interesting for birds. Most of the bird species that nest along rides or in glades are also typical of young plantations and young to middle-aged coppice. Where a wood has only small areas of these ephemeral open habitats, carefully designed and managed rides can help to enrich the bird life.

Shrubby margins along rides can be attractive to several species of breeding migrant birds – garden warbler, willow warbler, chiffchaff and nightingale. Low, tangled vegetation at the edges of rides is a particularly favoured nest site of chiffchaff. There is still much to be

Rob Fuller/BTO

This open glade with scattered mature trees and dead wood provides a range of wildlife habitats.

Bruce Henry/Frank Lane Picture Agency

Crumbling, sunny south-facing banks are valuable as nesting sites for bees and wasps.

Martin Warren

A 'box-junction' where rides cross (see p. 28) can be developed into a glade with a minimum of tree removal.

Eric & David Hosking

Nightingales may nest in shrubby margins of rides and glades.

Peter Wakely/NCC

This small glade in blackthorn scrub is the habitat of nightingales and the rare black hairstreak butterfly.

Heather Angel

Cutting down some of the trees and scrub is an essential part of ride management, but it should be done a section at a time over several years.

Peter Wakely/NCC

Spring flowers such as primroses flourish on recently cut ride edges.

D Avon/Ardea

The chiffchaff favours low, tangled ride-side vegetation for its nest.

E A Janes/NHPA

Wayleaves for electricity pylons or poles can be managed to benefit birds and butterflies.

Ian Beames/Ardea

The wood white butterfly prefers lightly shaded rides and soon disappears when neighbouring trees grow tall (see p. 13).

Ian Beames/Ardea

The white admiral lays its eggs on honeysuckle, usually in dappled shade.

Martin Warren

Scalloped clearings (see pp. 24-25) provide sheltered sunny areas that are favoured by many butterflies.

John Mason/Ardea

The grizzled skipper is one of the butterflies that require short vegetation in sunny areas for egg-laying (see p. 15).

learnt about ride management for birds but it seems that, to create suitable habitats for these breeding migrants, the shrubby strip between the open ride surface and the adjacent stand should not be less than 5 m wide and ideally should be much broader. One management option that could favour migrant birds is the cutting of coppice belts of 10 - 20 m either side of rides. Such shrubby strips and coppice are typically of greater value for breeding when they have 3 - 8 years of growth. Wrens are often associated with rides, particularly where occasional heaps of brushwood have been left. In woods lacking compartments of young trees, the distribution of tree pipits is usually closely tied to rides. It seems that they avoid narrow rides and that their territories are often positioned at the junction of several rides. In some large English conifer plantations, woodlarks will selectively use ride margins running alongside stands of very young trees in preference to those which run through mature stands. The nature of the adjacent crop may well influence the way that other species use rides, but present information is scanty.

Some bird species use rides for feeding whilst others are attracted to them for breeding. Rather few species feed on the open ride surface although blackbirds, song thrushes and robins nesting in the adjacent woodland frequently do so. Other species that feed along rides include sparrowhawks, which often hunt along them for small birds. The woodcock may make special use of rides and glades in its curious courting behaviour. At dawn and dusk the males go on their display flights – 'roding' above the woodland canopy. They pay close attention to rides and glades, presumably looking for females which have moved out of the closed woodland into the open areas. In summer, woodcock also use glades for night-time roosting.

The richest glades for breeding birds are those with a mosaic of open vegetation and scrub. Such areas can provide breeding habitats for a variety of species including warblers, tree pipits and yellowhammers. Natural woodland glades were probably once an important breeding habitat of the nightjar but nowadays it is rarely encountered in glades, being mainly confined in Britain to heaths and young plantations. Green woodpeckers are often attracted to glades, and also sometimes to rides, to feed on ants. In autumn and winter, thrushes can be attracted to those glades with patches of scrub, to feed on the berries of shrubs such as hawthorn. Very small glades, of less than about one quarter of a hectare, and those dominated purely by bracken are generally of little interest for birds.

Mammals

Ride-side grassland and coppice can support large populations of small mammals, which, in turn, can attract predators such as kestrels and owls. Where shrubs or coppice are cut on a cycle, numbers of small mammals appear to increase during the first three or four years after cutting and then decrease. Whilst rides are probably beneficial to most small mammals, they may create a problem for the common doormouse. This scarce species is largely arboreal, spending much of its life amongst the branches and foliage of shrubs and trees. Consequently, wide rides with no interconnecting tree canopy can be a formidable barrier to this attractive animal. Therefore, special care should be taken in planning the design and management of rides in woods known to hold common dormice. Occasional 'ride bridges', provided by canopy trees, shrubs or even single branches, may greatly aid the movement of dormice from one part of a wood to another. Hares are commonly found deep inside woodlands and may be dependent on grassy rides and glades as feeding sites. Rides can offer a good means of observing and, where necessary, controlling deer within woodland. Bats use rides for hunting; so, if bat boxes are thought desirable, they should be situated on large trees, facing out over rides or glades.

Rides and game

An assessment of the value of rides would not be complete without acknowledging their importance in woods managed for pheasant shooting. Rides provide positions for the guns, flushing points and suitable feeding sites. Moreover, pheasants are woodland-edge birds and the existence of open rides will draw pheasants inside woods where ride margins can offer suitable breeding sites. It is perfectly feasible to manage rides in such a way that they benefit both game and wildlife, and several of the management options described below would achieve this.

The management of woodland rides

Without management and in the absence of heavy grazing, scrub and trees will quickly encroach on any open area and its special wildlife will gradually disappear. Some form of management is needed, therefore, to preserve and even to enhance the conservation value of rides and glades. We stress, however, that before embarking on any management initiative it is essential to have clearly defined objectives. For example, is the management aiming to provide favourable conditions for one or a few particular species, or is it aiming to create as rich a wildlife as possible? There are three dominant features of a ride that have a major influence on the wildlife – the level of shade, the ride width and the type of cutting or grazing regime.

Shade and width

The level of shade in a woodland ride is affected by three main factors – the width of the ride, the height of the surrounding trees, and the orientation of the ride with respect to the sun. The interrelationship of these three factors is shown in Figure 3. In general terms, the wider the ride or the smaller the surrounding trees, the sunnier the ride will be. Also, rides orientated east/west will receive more sun during the summer months than those orientated north/south (although this is reversed during the winter months in narrower rides and also in the far north of Britain).

Many plants and animals, notably the majority of butterflies, occur only in very open, sunny rides with less than about 20% direct shade. However, certain species prefer partially-shaded or heavily-shaded rides, and a variety of shade conditions are needed to cater for a wide range of wildlife. Few woods have very open, sunny rides and most would benefit considerably from widening a few (but not all) rides, or the creation of some large glades. As a rule of thumb, the most open rides need to be at least as wide, and preferably 1.5 times as wide, as the height of the surrounding trees. Thus, in mature woodland, where the trees are 20 – 30 m tall, a ride will need to be at least 30 – 45 m wide to cater for the most sun-demanding species. This width is measured from the base of the mature or crop trees on either side, and substantial scrub or coppice margins can be present along the margins without overshadowing the ride. Indeed, wide shrubby margins are desirable for moths and breeding birds, as described above.

In plantations composed of stands of different ages, tree heights will vary considerably and a network of narrower rides (about 5 – 15 m wide) may be satisfactory. In such woods, ride orientation will also be less important. However, although narrow rides may be suitable for a wide range of species when the adjacent trees are small, they will eventually deteriorate as they become more shaded, and a few wide rides will always be beneficial. A scatter of taller, standard trees can provide additional habitats within the ride and often make the ride more visually attractive. They will not decrease light

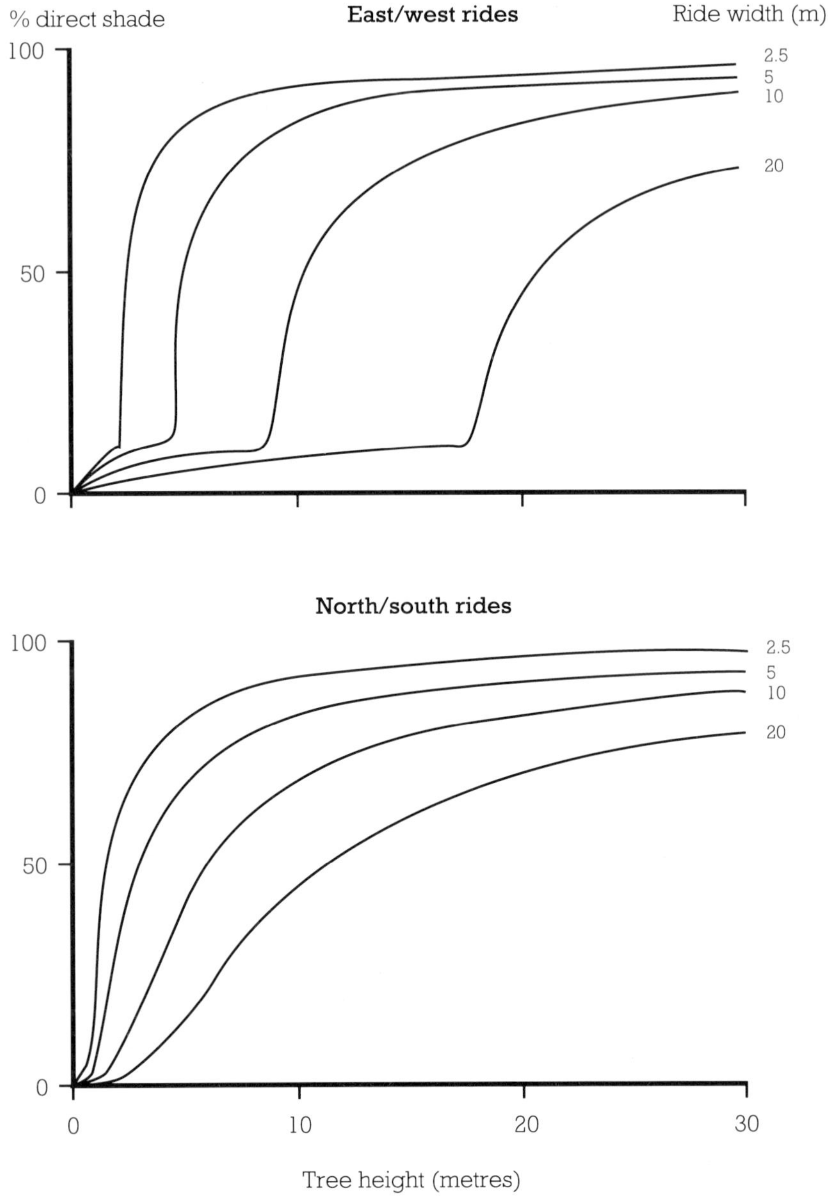

Figure 3 The relationship between the level of direct shade during June and tree height in woodland rides of different widths and orientation (calculated for latitude 52°N, which runs near Cambridge)(after Warren 1985)

levels substantially provided that they amount to less than about 10% canopy cover.

Cutting regimes

In the past, open habitats in woods were often maintained by grazing animals, but this is rarely practicable today and has largely been replaced by cutting. The type of cutting regime is crucial to wildlife, as it influences both the composition and the structure of the vegetation. Many invertebrates, for example, are restricted to one type of vegetation structure. Some species prefer bare ground, some short or sparse grassy vegetation, some taller grass and herbs, while others prefer various stages of scrub or coppice regrowth. To encourage rich communities of invertebrates the management should aim to create a variety of conditions. The vegetation structure is determined not only by the cutting regime, but also by inherent site factors such as soil type and climate. Therefore, the same regime may have different effects on different sites.

The least beneficial cutting regime is the regular mowing of a ride from edge to edge, as this eliminates all structural variation. The structure of the vegetation can be varied most easily by dividing the ride into strips running parallel to the ride edge and giving each strip a different treatment. Although the precise effects of different cutting regimes are still the subject of current research, four examples are described below with an assessment of their strengths and weaknesses. The most appropriate option will depend on the aims of management, the type of the wood, the resources or machinery available, the type of soil, and so on.

System 1: simple two-zone system

This comparatively simple system of ride management (Figure 4) is most appropriate to woods where costs are to be kept to an absolute minimum. It is an adaptation of the system already in operation in many commercial conifer forests. The central portion of the ride is mown once or twice annually to maintain

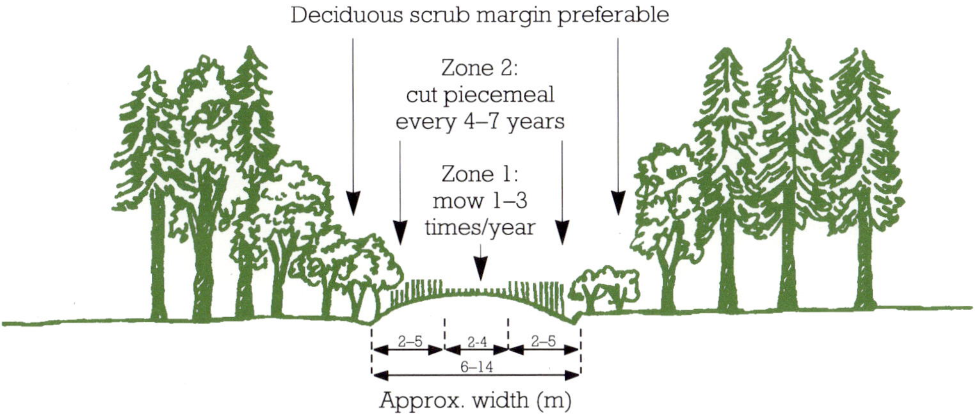

Figure 4 System 1: simple two-zone system

easy access, and the marginal strip either side is allowed to develop into a mixture of tall grass, herbs and light scrub. This is then cut on a 4 – 7 year rotation with a side-mounted flail. It is vital that these ride margins are cut in a piecemeal fashion, so that only a proportion is cut each year. This will create a moderate range of vegetation structure and will also ensure that there is a continuity of habitat for those birds and insects requiring tall grasses and herbs or young scrub. For example, many of the butterflies requiring open grassland can thrive under this system (see Table 1). Further variety can be provided by having some ride margins cut on a short (e.g. four-year) rotation, while others are cut on a longer rotation (e.g. every seven years). The main drawback of this system is that it does not permit the development of the dense scrub or coppice needed by many species of birds and insects.

System 2: three-zone system with coppice

This more complex system, depicted in Figure 5, should cater for a far greater range of species than System 1. It is probably most suitable for nature reserves, although it would undoubtedly be highly beneficial in any wood.

The central strip is again mown once or twice annually, but it is bordered on either side by a zone of tall grass and herbs. This zone is probably best cut piecemeal on a 3 – 4 year rotation with opposite sides of the ride being cut in different years. It is preferable to spread the work evenly over the years so that only one third or one quarter of the ride margins are cut each year. A simpler system of cutting this zone alternately every two years (that is one side one year, the other side the next) is also often used on nature reserves. Although this is

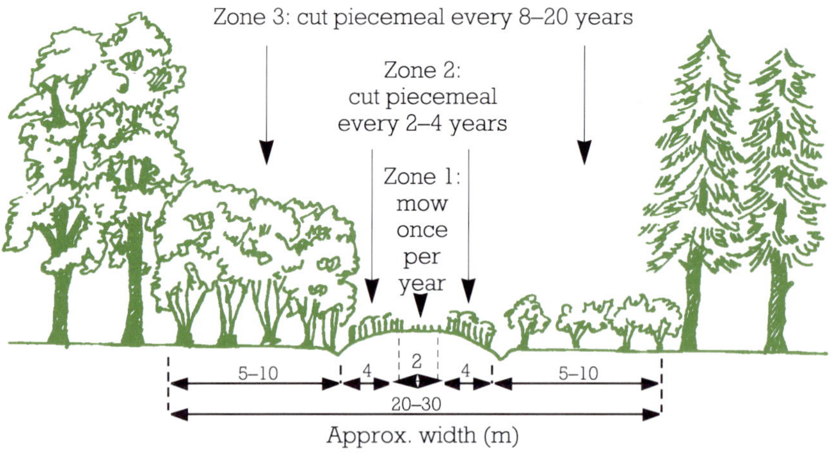

Figure 5 System 2: three-zone system

undoubtedly beneficial to many species, it may be rather too frequent for those insects preferring longer-established taller vegetation and infrequent disturbance (e.g. the wood white butterfly).

On either side of the strip of tall grass and herbs is a zone of scrub and tree regrowth, which is managed as coppice and cut on a rotation of 8 – 20 years. If there are insufficient scrub or coppice species already within this zone, it may be necessary to encourage natural regeneration by protecting any tree seedlings with tree shelters. Alternatively, the scrub zone can be planted or restocked with appropriate deciduous species. The coppice should also be cut in a piecemeal fashion, either by cutting 50 – 100 m lengths each year or by dividing the ride system into longer units (not necessarily of similar lengths) which can be cut in rotation.

Shorter coppice rotations (e.g. 8 – 10 years) are likely to be most suitable where there is a high density of scrub or coppice stools in the ride margins and where the regrowth is vigorous. Longer rotations (e.g. 15 – 20 years) are best when the regrowth is less vigorous or where a more mature crop is required, for example for firewood, the sale of which can help to offset the costs of ride management.

The main drawback with longer coppice rotations is that a smaller proportion of the ride margins will be coppiced each year, leading to more heavily shaded conditions. Although this reduces the labour required in management and possibly provides more revenue, it may not be the best option for wildlife. Many woodland species thrive only in freshly cleared areas and consequently would benefit from a shorter rotation.

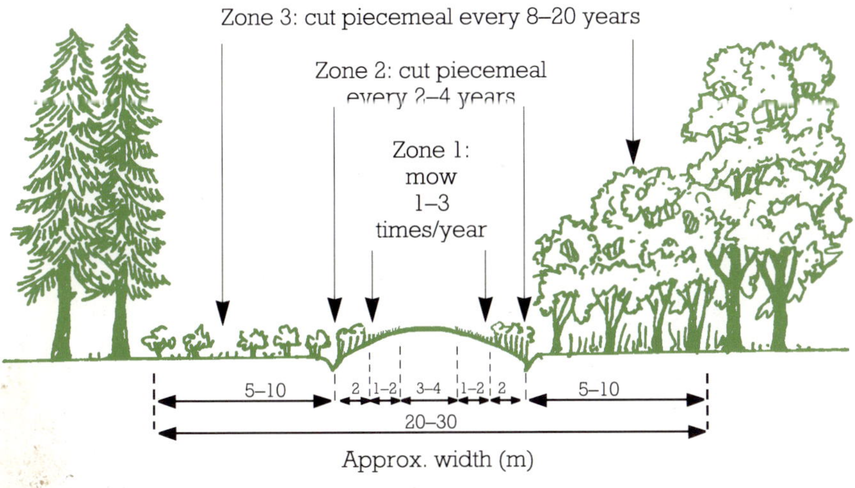

Figure 6 System 3: forest roads (three-zone system)

Species such as the pearl-bordered fritillary have declined severely in Britain now that active coppicing has ceased in most woods, but they can probably survive in suitably managed rides. If the conservation of such species is a priority and full-scale coppicing is not feasible (see Fuller & Warren 1990), then a short coppice rotation in the ride margins may be an excellent approach. In large woods with extensive networks of rides, it may be possible to vary the rotation length along different rides, perhaps in combination with other systems of management such as System 4. Indeed, such an option may be preferable to uniformly short coppice rotations because, within the wood as a whole, it will create a wider range of habitats for a larger number of species.

System 3: forest roads

Forest roads offer an excellent opportunity for encouraging the more light-demanding plants and animals. Roadways normally have to be wider than most woodland rides to prevent tree overhang, which restricts access by large vehicles and can cause damage through water dripping onto the road surface. With a minimum of extra widening such roadways can therefore be highly valuable for wildlife. The ideal management alongside forest roads is essentially similar to System 2, although many variations are possible such as those suggested under Systems 1 and 4. The main modification is that the short grass/herb zone is moved to a narrow strip on either side of the road surface (Figure 6). This zone is maintained either by annual cutting or by periodic grading to increase run-off. Either side of this strip is a zone of tall grass and herbs and a zone of scrub or coppice, both managed as in System 2.

System 4: scalloped edges

A potential problem with wide rides is the channelling of wind, which may adversely affect the sheltered conditions favoured by some woodland animals, notably insects. The true significance of the wind-

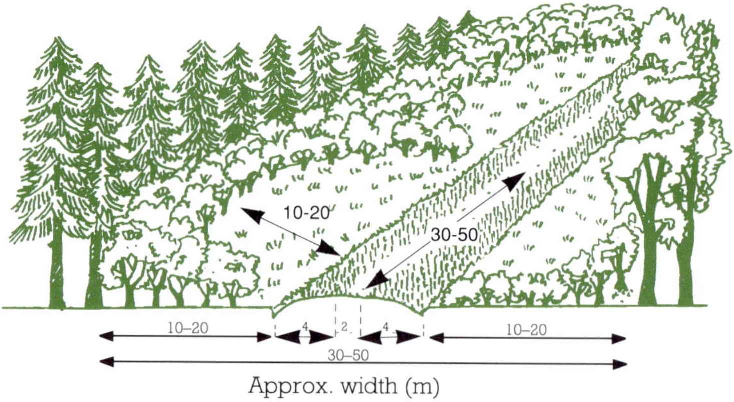

Figure 7 System 4: scalloped edges (three-zone system)

tunnel effect is, however, poorly understood. The problem is most acute in wide, straight rides and a simple solution is to have gently curving rides. In straight rides, shelter can be increased in several ways – either by creating wavy edges, by leaving a scatter of standard trees, or by scalloping the ride margins rather than widening them uniformly (Figure 7). Scalloping is also advantageous to wildlife as it increases the length of edge between different vegetation types. The ideal management here is to create a series of scallops along both sides of a ride and to manage them on some form of coppice rotation. Alternatively, some scallops could be cut more regularly and maintained as permanent grassy areas. The scallops can be created opposite each other to maximise the sunlight falling in the ride, or they can be staggered to create a series of sheltered bays. The most suitable size of each scallop is probably in the region of 30 – 50 m long and 10 – 20 m deep, although these dimensions are not critical. The central grassy portion of the ride can be managed as in either System 1 or 2.

When and how to cut

Unfortunately, little information is available about the best time for cutting woodland grassland. Cutting during the autumn (or occasionally winter) probably has the least direct effect on invertebrate populations. However, there is some evidence that cutting during the non-growing season may encourage a few vigorous grasses to dominate, leading to a severe reduction in plant diversity. This problem is potentially most serious on deeper, more fertile soils, particularly clay soils, which often become dominated by tufted hair-grass.

In the past, rides that were cropped for hay would have been cut in July or possibly August. Although cutting at this time may harm certain invertebrates, some form of summer cutting may actually be necessary to maintain a herb-rich sward in the long term, including the maintenance of numerous insect food-plants. It is also unlikely significantly to reduce populations of most invertebrates in the short term provided only a small proportion is cut each year, as suggested for zone 2 in Systems 1 – 4. A possible alternative would be to cut some rides during the spring or very early summer, which would have the advantage of removing early grass growth whilst still allowing many herbs to flower later in the season. Once again, some variation in the timing of cutting of different rides may be worth attempting, as different species will undoubtedly benefit from different regimes. Ride-edge coppice is best cut during the winter when the trees are dormant and when there is no danger of disturbing nesting birds or feeding insects.

Several problems may arise if too much cut material is left on the ride. First, the vegetation can be smothered, thereby reducing the regrowth or germination of less vigorous plants and ruining the habitat for certain insects, as well as for the plants themselves. Secondly, it may lead to problems associated with nutrient build-up in the long term.

Both these factors will tend to reduce the variety of plants, particularly if cut material is repeatedly left over a period of many years. It is, therefore, preferable to remove as much of the cut material as possible. This was done in the past, when rides were cut for hay, and this is known to have maintained herb-rich conditions. Today, most rides are cut by tractor-drawn cutters or swipes which leave all the cut material behind as a mulch. Ideally, this should be either raked and stacked into heaps or removed altogether. The heaps can provide an additional habitat in themselves for grass snakes and invertebrates such as spiders. This is a fairly labour-intensive process, however, and it may be cheaper to use a robust type of forage-harvester (e.g. Russell harvester) which can cut and gather the vegetation in a one-pass operation. In some conifer plantations, a forage-harvester could possibly be adapted to cut the ride vegetation and blow the cut material into the adjacent crop. This solution is unsatisfactory in deciduous woodland, however, as this tends to have a more interesting ground flora which could be damaged by the process.

Light marginal scrub can be cut either by a tractor with a side-mounted flail or by a hand-held clearing-saw. These will probably be suitable for scrub up to 5 – 7 years old. Thicker, older scrub or coppice will need to be cut with a chain-saw. The brash can either be dragged within the woodland itself or, preferably, burnt carefully in small heaps. To minimise the impact on the ground vegetation, it is best to keep the burn sites as small as possible and to use the same sites when recutting in future years. If neither of these options is practicable, the larger brash can be gathered together into stacks and left to rot down within the ride margin. Smaller brash will rot down much more quickly and can be left scattered over the ground.

A beneficial side effect of cutting woodland vegetation is that it often disturbs the soil, which can encourage plant germination. Light rutting of rides by vehicles and shallow mechanical grading to level the ride surface can increase the variety of plants, but only if the disturbance is irregular and not too severe. Digging out ditches to maintain good ride drainage can be equally important, not only for encouraging plant germination but also for providing bare ground habitat, which is vital for certain insects, notably bees and wasps. All these mechanical operations can, therefore, be of some benefit to wildlife, provided that they affect only part of the ride margins in any given year.

Creating new rides

The wildlife of many woods can be enhanced considerably by the creation of new rides or the widening of existing ones, following the guidelines given above. However, it should always be borne in mind that rides are a relatively recent development within most woods, and some semi-natural woods with no existing ride networks may best be left in an untouched state. This is particularly true for surviving wood-pasture systems and for woods with a very rich ground flora. The final decision will rest on the aims of management within each wood. New rides can be created by felling all trees along the proposed route, perhaps leaving a scatter of standards or a few groups of standards here and there. The stumps in the centre of the ride, which will become the grass/herb zones, can be treated with a herbicide such as Garlon, Amcide or Roundup. The stumps along the margins should not be treated, but allowed to regrow, subsequently being managed as coppice, as for example under one of Systems 1 – 4 described above. The lay-out of a new ride system will have to take account of practical site factors such as access points and slope, in order to be useful for timber extraction. Consider creating gently curving rides to reduce the wind-tunnel effect.

Ride widening – a word of caution

The benefits of ride widening are undoubtedly considerable in many woods, but there may sometimes be a risk of creating mediocre grassland at the expense of valuable semi-natural woodland. The process of ride widening can also damage certain forms of wildlife if too much is done too quickly. The danger is greatest in conifer plantations, where often the only remaining deciduous trees are found along ride edges or stream sides. An extensive programme of ride widening can remove much of this deciduous scrub zone, which may be the only habitat in the wood for numerous animals. The solution is to proceed with caution and do a little each year. If certain areas need to be widened quickly, try to retain substantial patches of ride-edge scrub and leave a number of rides untouched. Also, try to keep specimens or patches of particular trees which are poorly represented in the wood as a whole. A patch of aspens may, for example, be the only breeding area for several species of moths. In addition, ride widening can occasionally lead to problems with windthrow when wind-hardened trees along the ride edges are removed, exposing weaker trees behind. The problem is rarely acute in lowland areas of Britain, but it may be serious in some upland plantations. It can be overcome easily in plantations either by widening rides when the trees are still fairly young or by widening rides in stages or by scalloping the ride margins.

The management of woodland glades

The presence of glades increases the variety of habitats for wildlife in a woodland. The creation of new glades may not be desirable in all woodland, because it can destroy important features in semi-natural woods. In each site, the aims of management should be clearly defined, taking account of the existing interest. In particular, certain species need fairly large open areas within woods and can be catered for more easily in glades than in rides. For example, the marsh fritillary rarely breeds in woodland rides (even though its food-plant – devil's-bit scabious – may often be abundant) and needs damp, grassy glades of at least one hectare (2 – 3 acres) in size, and preferably several of them. Such glades can be of additional value as deer lawns or for game interest.

Even small openings, or bays, along the edges of rides will benefit some species. However, to encourage a variety of wildlife, some glades should be at least 0.25 hectare (0.5 acre) and preferably 0.5 – 2 hectares (1 – 5 acres) in size. A substantial glade can be created most easily by enlarging the intersection where two rides cross (or a T-junction) to create an intersection glade or 'box-junction' (Figures 8 and 9). This creates the maximum amount of glade for the minimum of tree removal. Alternatively, glades can be made within the woodland adjacent to a ride, linking two rides or completely separated from the ride network.

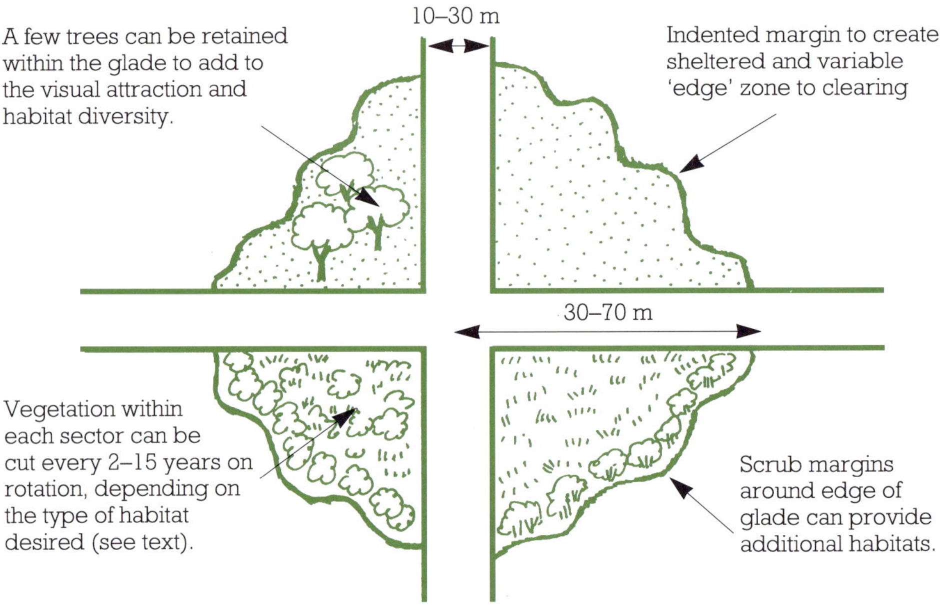

Figure 8 Diagram of a 'box-junction' clearing at the intersection of two rides

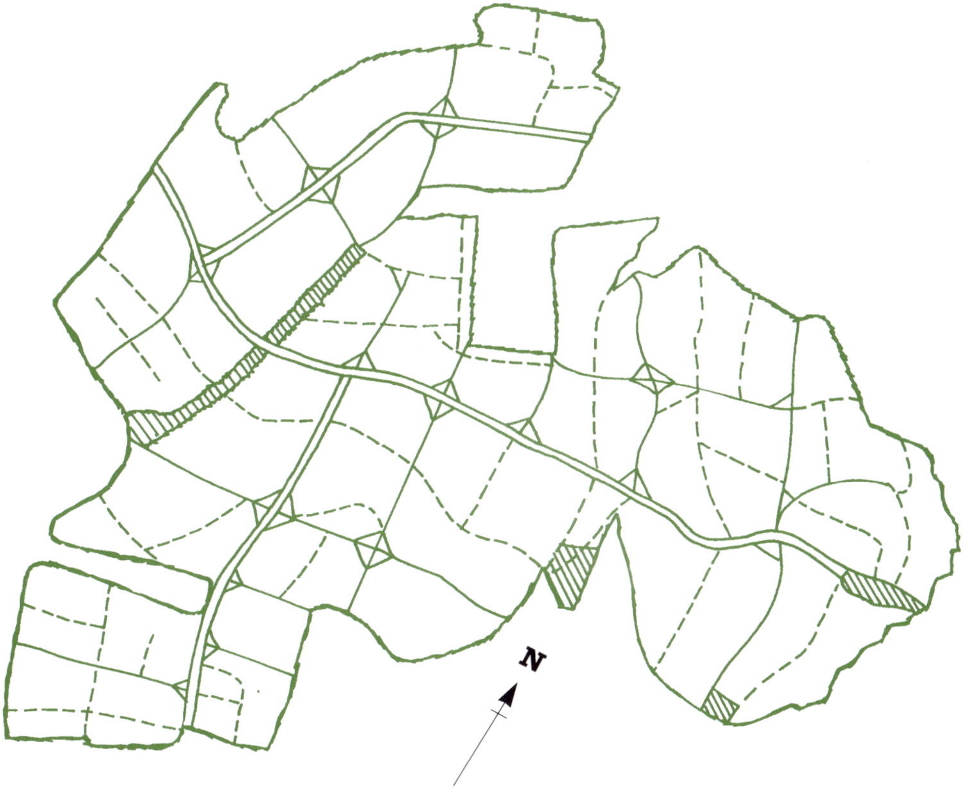

Figure 9 Bernwood Forest Nature Reserve, Oxfordshire (Shabbington Wood): a Forestry Commission wood managed for both timber and wildlife. The wood is particularly important for butterflies and moths, many of which breed in the rides and glades. The management agreement with English Nature provides for a network of interconnecting rides, both wide and narrow, with glades at ride intersections. The rides and glades together constitute 1% of the forest area.

0 0.5 1 km

- Forest roads (at least 15 m wide)
- Major rides (at least 15 m wide)
- Minor rides (less than 15 m wide)
- Box or T–junction glades
- Special management areas (scrub)

The principles of management within glades are essentially the same as those described for rides, and many different systems are possible. It is preferable to provide a range of vegetation types, varying from open grassland (with short and tall grasses and herbs) through to light and dense scrub. A scatter of taller trees will also be of considerable benefit to wildlife, as well as enhancing the visual appearance. Vegetation structure can be varied by having patches of grassland cut at different frequencies (e.g. every 1 – 4 years) at different times of the year. Patches of bramble, bracken or light scrub should be cut less frequently (e.g. every 3 – 6 years) and denser scrub every 8 – 20 years in a piecemeal manner. A system of irregular cutting within patches of scrub can provide a good mosaic of habitats. The management of the vegetation in box junctions can also be varied to suit different species. For example, each of the four sectors of the junction could receive a different cutting regime so that some sectors are cut annually or every 2 – 3 years (to create more grass and herb-dominated vegetation) while others are cut on longer rotations of 4 – 15 years (to allow the development of scrub- and tree-dominated vegetation).

Control of grazing is more practicable in woodland glades than in rides. Indeed, this was often the traditional management in glades. Grazing creates a different type of vegetation from cutting and therefore may benefit a rather different range of plants and animals. As with cutting, the type of grazing regime can be varied to suit different wildlife species, and the precise effects of each regime will vary from site to site. Often relatively light grazing will be most beneficial to plants and insects, as it creates diverse vegetation while preventing the invasion of scrub and trees. Elsewhere, a form of rotational grazing may be preferable, whereby stock are grazed within a small, fenced area (for example by using electric flexi-netting) and then moved periodically either around a single large glade or from one glade to another. In the latter approach, the principle is to graze each area reasonably heavily for a short period (but avoiding excessive poaching of the ground), followed by a longer period without grazing, to create a mosaic of open vegetation with different turf heights.

Acknowledgements

This booklet resulted from research programmes now commissioned by JNCC, both in-house and from the British Trust for Ornithology, the Institute for Terrestrial Ecology and other ecological consultants. We thank the following who have kindly contributed towards this booklet by commenting on earlier drafts or by supplying information - John Bratton, Ray Collier, Steven Falk, Nick Greatorex-Davies, Rhys Green, Andrew Henderson, Graham Hirons, Dick Hornby, Roger Key, Keith Kirby, Ian McLean, Maurice Massey, Keith Payne, George Peterken, Mike Pienkowski, Caroline Steel, David Steel, Alan Stubbs, Mark Tasker and Paul Waring. Stefa Birkenhead and Philip Oswald edited the text for publication.

Further reading

BRATTON, J.H., & ANDREWS, J. 1991. Invertebrate conservation – principles and their application to broad-leaved woodland. *British Wildlife, 2,* 335-344.

BRIGHT, P.W., & MORRIS, P.A. 1989. *A practical guide to dormouse conservation.* Occasional publication of the Mammal Society, No. 11. London, The Mammal Society.

BRITISH TRUST FOR CONSERVATION VOLUNTEERS. 1980. *Woodlands: a practical conservation handbook.* Wallingford.

CARTER, C.I., & ANDERSON, M.A. 1987. *Enhancement of lowland forest ridesides and roadsides to benefit wild plants and butterflies.* Farnham, Forestry Commission. (Research Information Note 126.)

COOKE, A.S. 1986. *The use of herbicides on nature reserves.* Peterborough, Nature Conservancy Council. (Focus on nature conservation No. 14.)

DENT, S., & SPELLERBERG, I.F. 1987. Habitats of the lizards *Lacerta agilis* and *Lacerta vivipara* on forest ride verges in Britain. *Biological Conservation, 42*, 273-286.

EVANS, J. 1984. *Silviculture of broadleaved woodland.* London, HMSO. (Forestry Commission Bulletin No. 62.)

FERRIS-KAAN, R., *ed.* 1991. *Edge management in woodlands.* Farnham, Forestry Commission. (Forestry Commission Occasional Paper, No. 28.)

FRY, R., & LONSDALE, D., *eds.* 1991. *Habitat conservation for insects – a neglected green issue.* Feltham, Amateur Entomologists' Society. (The Amateur Entomologist, *21.*)

FULLER, R.J., & WARREN, M.S. 1993. *Coppiced woodlands: their management for wildlife.* 2nd ed. Peterborough, Joint Nature Conservation Committee.

FULLER, R.J., & WARREN, M.S. 1991. Conservation management in ancient and modern woodlands: responses of fauna to edges and rotations. *In: The scientific management of temperate communities for conservation,* ed. by I.F. Spellerberg, F.B. Goldsmith and M.G. Morris, 445-471. Oxford, Blackwell Scientific Publications. (British Ecological Society Symposium No. 31.)

FULLER, R.J., & WHITTINGTON, P.A. 1988. Breeding bird distribution within Lincolnshire ash-lime woodlands: the influence of rides and the woodland edge. *Acta Oecologica/ Oecologia Generalis, 8,* 259-268.

HILL, D.A., & ROBERTSON, P.A. 1988. *The pheasant: ecology, management and conservation.* Oxford, Blackwell Scientific Publications.

KIRBY, K.J. 1984. *Forestry operations and broadleaf woodland conservation.* Peterborough, Nature Conservancy Council. (Focus on nature conservation No. 8.)

KIRBY, P. 1992. *Habitat management for invertebrates: a practical handbook.* Sandy, Royal Society for the Protection of Birds, for the Joint Nature Conservation Committee and National Power.

PETERKEN, G.F. 1981. *Woodland conservation and management.* London, Chapman and Hall.

RACKHAM, O. 1980. *Ancient woodland: its history, vegetation and uses in England.* London, Arnold.

WARREN, M.S. 1985. The influence of shade on butterfly numbers in woodland rides, with special reference to the wood white, *Leptidea sinapis. Biological Conservation, 33,* 147-164.